Fast Facts About
BUTTERFLIES

by Lisa J. Amstutz

PEBBLE
a capstone imprint

Pebble Emerge is published by Pebble, an imprint of Capstone.
1710 Roe Crest Drive, North Mankato, Minnesota 56003
www.capstonepub.com

Library of Congress Cataloging-in-Publication Data
Names: Amstutz, Lisa J., author.
Title: Fast facts about butterflies / by Lisa J. Amstutz.
Description: North Mankato, MN : Pebble, an imprint of Capstone, [2021] | Series: Fast facts about bugs & spiders | Includes bibliographical references and index. | Audience: Ages 6–8 | Audience: Grades 2–3 | Summary: "Flitter and flutter! It's a pretty butterfly! Young readers will get the fast facts on these colorful insects, including butterfly body parts, habitats, life cycles, and why they are important to the environment. Along the way, they will also uncover surprising and fascinating facts! Simple text, close-up photos, and a fun activity make this a perfect introduction to the beautiful world of butterflies." —Provided by publisher.
Identifiers: LCCN 2020031934 (print) | LCCN 2020031935 (ebook) | ISBN 9781977131492 (hardcover) | ISBN 9781977132666 (paperback) | ISBN 9781977154170 (pdf) | ISBN 9781977155887 (kindle edition)
Subjects: LCSH: Butterflies—Juvenile literature.
Classification: LCC QL544.2 .A486 2021 (print) | LCC QL544.2 (ebook) | DDC 595.78/9—dc23
LC record available at https://lccn.loc.gov/2020031934
LC ebook record available at https://lccn.loc.gov/2020031935

Image Credits
Shutterstock: Brian Lasenby, 6, Cathy Keifer, 16, 17, Cornel Constantin, 10, Eric Isselee, cover, Gokula Priya Eswaran, 9, Inspiration GP, 20 (left), jakrit yuenprakhon, 14, JHVEPhoto, 19, jonathan_law, 5, Ken Griffiths, 13, liu yangjun, 20 (clothespin), Malgorzata Wryk-Igras, 21, marla dawn studio, 20 (bottom middle), Mega Pixel, 20 (top right), Peaw_GT, 11, Petr Ganaj, 7, Raafi Nur Ali, 15, riphoto3, 20 (bottom right), Russell Marshall, 8, SweetLemons, 20 (eyes), Wansfordphoto, 18, zabavina (background), cover and throughout

Editorial Credits
Editor: Abby Huff; Designer: Kyle Grenz; Media Researcher: Jo Miller; Production Specialist: Tori Abraham

All internet sites appearing in back matter were available and accurate when this book was sent to press.

Printed and bound in the USA. PO 3837

Table of Contents

Words in **bold** are in the glossary.

All About Butterflies

Flap, flap! A butterfly flies by. It lands on a flower. Can you count its legs? It has six. It is an **insect**. It also has three body sections. See the two long **antennae** on its head? They can sniff out food. Each one has a bulb at the end.

wings

antennae

legs

5

A butterfly has four wings. The wings have tiny scales. These pieces shine when light hits them. They give the wings color.

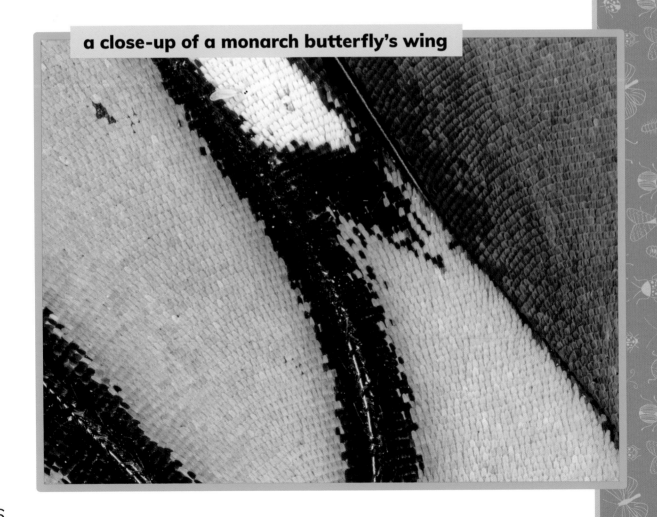

a close-up of a monarch butterfly's wing

flies come in many colors. Some

in with plants. This helps butterflies

Others have bright colors or big spots.

s warns enemies to stay away.

Queen Alexandra's birdwing is the largest butterfly in the world.

There are more than 17,000 kinds of butterflies. They can be as big as a dinner plate! They can be as small as your fingernail.

Butterflies live all over the world. They live in forests. They live in hot deserts and on cold mountains. But they do not live in Antarctica. It is too cold there!

Time to Eat

A butterfly's mouth is like a straw. It is very long. It curls up when it's not in use. The insect stretches it out to drink.

Many butterflies drink **nectar**. It is a sweet liquid from flowers. They also drink from mud puddles. They get salts and **minerals** there. This keeps them heathy.

Butterflies help plants. They **pollinate** flowers. How? A butterfly lands on a flower. It reaches in for nectar. A powder from the flower sticks to its body. This is pollen. The insect goes to a new flower. The pollen brushes off. Now fruit and seeds can grow.

A Butterfly's Life

A female butterfly lands on a leaf. She uses her feet to taste it. Is it the right kind of plant for her young to eat? Yes! She lays her eggs here.

egg

A **larva** hatches out of each egg. It is also called a caterpillar. It eats the leaf. It grows. Soon it sheds its skin, or **molts**. It will molt four or more times. Each time it gets bigger.

Finally, the larva makes a case around itself. The case is called a **pupa** or chrysalis. Inside, the bug's body changes. It may take a few weeks. Some stay inside for years.

a larva making a pupa

a butterfly leaving the pupa

The butterfly is ready. It breaks out of the case. Its wings are wet. It slowly flaps them. The wings dry. Now it can fly. Bye, butterfly!

Fun Facts

- Butterflies and moths are related, but they are not the same. Moth antennae are feathery. They feed at night. Butterflies eat during the day.

- The glasswing butterfly has clear wings. You can see through them!

glasswing butterfly

monarch butterflies in Mexico during winter

- The monarch butterfly flies south in winter. It can travel up to 3,000 miles (4,828 kilometers).

- Some butterflies feed on animal poop!

Build a Butterfly

What You Need:

- markers or watercolor paints
- coffee filter
- clothespin
- half a pipe cleaner
- two googly eyes
- glue

What You Do:

1. Draw or paint a pattern on the coffee filter. This will make the wings. Let it dry.

2. Pinch the middle of the filter together. Clamp it with the clothespin.

3. Bend the pipe cleaner into a V. Put it in the clothespin to make antennae.

4. Glue the googly eyes to the clothespin.

Glossary

antenna (an-TEH-nuh)—a feeler on an insect's head used to touch and smell

insect (IN-sekt)—a small animal with a hard outer shell, six legs, three body sections, and two antennae

larva (LAR-vuh)—an insect at the stage of its life cycle between an egg and a pupa; a butterfly larva is also called a caterpillar

mineral (MIN-ur-uhl)—a material found in nature that is not made by an animal or a plant

molt (MOHLT)—to shed an outer layer of skin

nectar (NEK-tur)—a sweet liquid found in many flowers

pollinate (POL-uh-nayt)—to move pollen from flower to flower; pollination helps flowers make seeds

pupa (PYOO-puh)—an insect at the stage of its life cycle between a larva and an adult; a butterfly pupa is also called a chrysalis

Read More

Dickmann, Nancy. *Butterfly.* Tucson: Brown Bear Books, 2020.

Dunn, Mary R. *A Butterfly's Life Cycle.* North Mankato, MN: Capstone Press, 2018.

Zemlicka, Shannon. *The Story of a Butterfly: It Starts with a Caterpillar.* Minneapolis: Lerner, 2020.

Internet Sites

Butterflies
ducksters.com/animals/butterfly.php

Butterflies and Moths
dkfindout.com/us/animals-and-nature/insects/butterflies-and-moths/

The Butterfly Life Cycle!
natgeokids.com/za/discover/animals/insects/butterfly-life-cycle/

Index